RASPBERRY PI 5 101

A beginner's guide to cracking the code and unleashing the magical powers of Raspberry Pi 5, effortlessly.

Andy Stewart

All rights reserved © Andy Stewart

INTRODUCTION

WHAT IS RASPBERRY PI 5?

Raspberry Pi 5 is a single-board computer, essentially a compact and affordable computer packed onto a circuit board. It's the latest addition to the popular Raspberry Pi line, known for its versatility and accessibility in the world of electronics and programming. Here's a breakdown of key things to know about Raspberry Pi 5:

Features:

Powerful Processor: Boasting a custom-designed Broadcom BCM2712 SoC with quad-core ARM Cortex-A76

CPUs clocked at 2.4GHz, it offers a significant performance boost compared to its predecessors, making it capable of handling more demanding tasks.

Enhanced Graphics: With dual 4Kp60 display outputs via micro HDMI, it delivers smooth visuals for browsing, watching videos, or even casual gaming.

Improved Connectivity: It comes equipped with Gigabit Ethernet and Wi-Fi 6, providing faster and more reliable internet connections.

USB-C Power: It utilizes a USB-C port for power, aligning with modern standards and offering improved power delivery.

Expandability: The familiar GPIO pins are retained, allowing you to connect sensors, LEDs, and other devices for electronics projects.

What can you do with Raspberry Pi 5?

Multimedia Center: Use it to stream movies, music, and TV shows, or even set up a retro gaming console with emulators.

Learning to Code: Its beginner-friendly operating system and programming languages like Python make it a great platform to learn coding.

Electronics Projects: With its GPIO pins and compatibility with various sensors and components, you can build robots, weather stations, smart home devices, and more.

Home Automation and IoT: Integrate Raspberry Pi into your smart home to control lights, thermostats, and other devices.

Server and Networking:. Its capabilities allow you to set up a web server, host files, or manage a network.

Benefits of Raspberry Pi 5:

Affordable: With a starting price of around $80, it's an accessible tool for learning and hobbyists.

Open-source: The operating system and many resources are free and open-source, fostering a large and supportive community.

Versatility: Its wide range of applications, from learning to code to building electronics, makes it a valuable tool for various interests.

Compactness: Its small size makes it easy to store and integrate into projects.

Whether you're a beginner exploring technology or an experienced maker, Raspberry Pi 5 offers a powerful and user-friendly platform for exploring your creativity and technical skills.

Understanding its place in the world of computers

Understanding Raspberry Pi 5's Place in the World of Computers:

Raspberry Pi 5 occupies a unique niche in the computer world, bridging the gap between several categories:

1. Educational Platform:

- Affordable and accessible: The ~$80 price tag makes it readily available for students and hobbyists compared to traditional desktops or laptops.
- Open-source software: Free operating systems like Raspberry Pi OS and extensive online resources foster learning and experimentation.

- Programming friendly: Easy-to-learn languages like Python and Scratch make it a perfect platform to grasp coding fundamentals.
- Electronics integration: GPIO pins and compatibility with sensors and actuators support hands-on learning in electronics and robotics.

2. Hobbyist and Maker Tool:

- Versatility: It offers endless possibilities for building projects, from multimedia centers and retro gaming consoles to home automation systems and robots.
- Community and support: A large and vibrant community provides inspiration, tutorials, and

troubleshooting assistance for diverse projects.
- Compact and portable: Its small size makes it ideal for prototyping and integrating into compact projects.
- Hackability and customization: With open-source hardware and software, the possibilities for tinkering and personalizing are vast.

3. Single-board Computer (SBC):

- Powerful enough for everyday tasks: Handles browsing, office applications, and even light gaming effortlessly.
- Energy-efficient alternative: Consumes significantly less power than traditional PCs,

making it eco-friendly and cost-effective.
- Headless server potential: Can be configured as a headless server for file sharing, web hosting, or network management.
- IoT gateway: Connects sensors and actuators to the internet for smart home and industrial applications.

4. Bridge between low-level and high-level computing:

- Hardware access: GPIO pins offer direct control over electronics, unlike most computers, promoting understanding of hardware-software interactions.
- Software development foundation: Provides a platform

to learn coding principles and build prototypes before transferring skills to more powerful computers.
- Introduction to embedded systems: Can be used to learn about embedded system programming and microcontroller applications.

In conclusion, Raspberry Pi 5 doesn't directly compete with traditional PCs or powerful servers. It fills a unique gap, offering an affordable, versatile, and accessible platform for learning, tinkering, and building projects that spark creativity and technical skills. It's a gateway to the world of electronics and coding, empowering individuals to explore their ideas and contribute to the broader field of technology.

Key features and capabilities overview

Here's a rundown of the key features and capabilities of Raspberry Pi 5:

1. Processing Power:

Quad-core ARM Cortex-A76 CPU: 2.4GHz clock speed offers significant performance boost over previous models, handling demanding tasks like 4K video editing and web development smoothly.

512KB L2 cache per core and 2MB shared L3 cache: Improves memory access and application responsiveness.

2. Enhanced Graphics:

VideoCore VII GPU: Supports OpenGL ES 3.1 and Vulkan 1.2, delivering improved graphics performance for gaming and media playback.

Dual 4Kp60 HDMI outputs: Connects to two 4K displays simultaneously for enhanced desktop experience or gaming setups.

3. Improved Connectivity:

Gigabit Ethernet: Provides high-speed wired network connection for fast internet access and file transfers.

Wi-Fi 6 (802.11ax): Enjoys faster wireless speeds, improved stability, and lower latency for seamless streaming and online gaming.

Bluetooth 5.0 and BLE (Bluetooth Low Energy): Connects seamlessly to various wireless devices like headphones, speakers, and IoT sensors.

4. Modern Interface:

USB-C power: Simplifies power delivery with the modern USB-C standard.

2x USB 3.0 ports: Offers high-speed data transfer for external storage devices.

2x USB 2.0 ports: Maintain compatibility with older peripherals.

5. Expandability and Flexibility:

40-pin GPIO header: Connects to sensors, LEDs, buttons, and other electronics for building projects and robotics applications.

MicroSD card slot: Supports high-speed SDR104 mode for faster boot times and data access.

PCIe 2.0 x1 interface: Enables connection to high-speed peripherals for future expansion.

6. Open-source Software:

Raspberry Pi OS: A user-friendly and free operating system based on Debian, with various software options pre-installed.

Extensive online resources: A vast community provides tutorials, projects, and support for beginners and advanced users.

7. Compact and Affordable:

Small form factor: Easily fits into projects and takes up minimal space.

Starting price of around $80: Makes it an accessible tool for learning and hobbyists.

Raspberry Pi 5 combines powerful processing, advanced graphics, improved connectivity, modern interfaces, and expandability at an affordable price. It's a versatile platform ideal for:

- Learning to code and electronics
- Building hobby projects and robots
- Creating a multimedia center or retro gaming console
- Setting up a home automation system

- Developing applications and exploring embedded systems

With its capabilities and accessibility, Raspberry Pi 5 opens doors to creativity and innovation, making it a valuable tool for beginners and experienced makers alike.

Comparing it to previous Raspberry Pi models

Comparing Raspberry Pi 5 to Previous Models: A Performance Leap

Let's dive into how Raspberry Pi 5 stacks up against its predecessors, particularly the highly popular Raspberry Pi 4:

Performance

CPU: Raspberry Pi 5 boasts a significant jump with its quad-core ARM Cortex-A76 CPU clocked at 2.4GHz, translating to roughly 2-3x faster performance compared to the Raspberry Pi 4's quad-core Cortex-A72 at 1.5GHz. This boost shines in demanding tasks like video editing, code compilation, and high-resolution gaming.

GPU:

The VideoCore VII GPU in Raspberry Pi 5 also delivers a considerable upgrade, supporting OpenGL ES 3.1 and Vulkan 1.2. This translates to smoother visuals, improved graphical processing for applications like games and media playback, and even potential for light AI workloads.

Memory:

Both Pi 4 and Pi 5 offer DDR4 memory options, but Pi 5 moves on to DDR4X. This newer generation boasts faster clock speeds and improved efficiency, contributing to the overall performance increase.

Connectivity:

Raspberry Pi 5 wins in the network battle with Gigabit Ethernet and Wi-Fi 6 (802.11ax), guaranteeing significantly faster internet speeds and more reliable wireless connectivity compared to the Pi 4's Gigabit Ethernet and Wi-Fi 5 (802.11ac).

Other Considerations:

Raspberry Pi 5 requires a new, more powerful power supply to unlock its full potential, adding a slight cost factor.

While both retain the familiar GPIO pins for electronics projects, Pi 5 utilizes a different I/O controller chip, potentially affecting compatibility with some older expansion boards designed for Pi 4.

Raspberry Pi 5 represents a significant upgrade over the Raspberry Pi 4, particularly in terms of processing power, graphics capabilities, and network connectivity. This makes it the ideal choice for users seeking the best performance and future-proofing. However, if cost is a primary concern and your tasks don't demand the additional power, the Pi 4 remains a capable and affordable option.

Ultimately, the choice depends on your specific needs and budget. If you're looking for the ultimate Raspberry Pi experience for demanding tasks and future-proof performance, Raspberry Pi 5 is the clear winner. If you're on a tighter budget or your needs are less intensive, Pi 4 may still be a suitable choice.

CHAPTER TWO

Unboxing and Setting Up Your Raspberry Pi 5

Identifying essential Components and essentials

Exciting times! You've got your Raspberry Pi 5, ready to embark on coding adventures, build interactive projects, or create a multimedia powerhouse. Now, let's navigate the unboxing and setup process together.

Essential Components:

1. Raspberry Pi 5 board: The star of the show! Be careful handling it, holding it by the edges to avoid touching sensitive components.

2. MicroSD card with Raspberry Pi OS pre-installed: This acts as your computer's hard drive and operating system. Choose a good-quality card like SanDisk or Samsung, with at least 16GB of storage for basic use.

3. USB-C power supply: You'll need a power supply capable of delivering at least 5V/3A. Raspberry Pi recommends an official 5.1V/3A power supply for optimal performance.

4. HDMI cable: Connect your Raspberry Pi to a monitor or TV. A micro HDMI to HDMI cable is required.

5. Keyboard and mouse: Navigate the operating system and interact with your Raspberry Pi. Any USB keyboard and mouse will work

Optional but Recommended:

Case: Protects your Raspberry Pi and improves heat dissipation. Consider an official case or explore various third-party options.

Micro HDMI to HDMI adapter: If your monitor/TV only has standard HDMI ports, you'll need an adapter.

Getting Started:

1. Prepare your microSD card: If your card doesn't come pre-installed with Raspberry Pi OS, download the latest version and flash it onto the card using a tool like Etcher.

2. Connect everything up: Insert the microSD card into the slot on the Raspberry Pi. Connect the HDMI

cable, power supply, keyboard, and mouse.

3. Power on! Press the power button on the Raspberry Pi. You should see lights come on and the screen display the boot process.

4. Welcome Wizard: Follow the on-screen instructions to set up your language, Wi-Fi connection, username/password, and time zone.

5. Explore and have fun! Raspberry Pi OS comes with various pre-installed applications for browsing, media playback, coding, and more. You can also install additional software and customize your experience.

Tips:

- Use a high-quality microSD card for better performance and reliability.
- Download the latest version of Raspberry Pi OS for bug fixes and new features.
- Check the official Raspberry Pi website for detailed setup guides and troubleshooting tips.
- Join the Raspberry Pi community forums for support and inspiration.

With these steps and essential components, you're well on your way to exploring the limitless possibilities of Raspberry Pi 5! Remember, the journey is just as exciting as the destination, so have fun tinkering, learning, and creating!

Connecting peripherals and power supply

Connecting peripherals and the power supply to your Raspberry Pi 5 is a straightforward process, but let's walk through it step-by-step for ultimate clarity:

1. Power Supply:

- Recommended: Use the official Raspberry Pi 5 Power Supply (5.1V/3A) for optimal performance and stability. Alternatively, a good-quality 5V/3A USB-C power supply from a trusted brand should work, but check Raspberry Pi's website for compatibility lists.
- Connecting: Locate the USB-C port on the side of the Raspberry

Pi 5. Gently plug the USB-C power cable into the port. Ensure the connection is secure.

2. HDMI Cable:

- Type: You'll need a micro HDMI to HDMI cable to connect your Pi 5 to a monitor or TV.
- Connecting: Locate the micro HDMI port on the side of the Raspberry Pi 5. Carefully plug the micro HDMI connector into the port. On the other end, connect the standard HDMI connector to your monitor or TV's HDMI port.

3. Keyboard and Mouse:

- Compatibility: Any USB keyboard and mouse will work. Wireless

options are convenient but require batteries or charging.
- Connecting: Locate the USB ports on the top and side of the Raspberry Pi 5. Plug the keyboard and mouse USB connectors into any available USB port.

4. Optional Peripherals:

- MicroSD Card: If your microSD card isn't pre-installed with Raspberry Pi OS, insert it into the slot on the underside of the Raspberry Pi 5.
- Ethernet Cable (Optional): If you prefer a wired internet connection, plug an Ethernet cable (RJ45) into the Ethernet port on the side of the Pi 5.

- Other Peripherals: You can connect additional peripherals like external hard drives, cameras, and sensors through the USB ports or GPIO pins (advanced users).

Tips:

- Ensure all connections are secure before powering on.
- If using an unofficial power supply, verify its compatibility with the Raspberry Pi 5.
- For troubleshooting, check the official Raspberry Pi website for FAQs and guides.

Remember: With everything connected, you're ready to power on your Raspberry Pi 5 and embark on your exciting tech adventures!

Setting up microSD card with operating system

Setting up your microSD card with Raspberry Pi OS for your Raspberry Pi 5 can be done in a few simple steps, whether you have a pre-installed card or need to flash it yourself. Here's a breakdown for both scenarios:

Option 1: Pre-installed Card:

1. Verify Raspberry Pi OS version: Check the label or packaging of your microSD card to ensure it has the latest Raspberry Pi OS version (currently December 2023).

2. Insert the card: Gently push the microSD card into the slot on the underside of your Raspberry Pi 5, with the gold contacts facing the board.

3. Power on: Press the power button on your Pi 5. You should see the green LED light up and the boot process begin.

4. Follow the welcome wizard: On-screen instructions will guide you through setting up your language, Wi-Fi connection, username/password, and time zone.

5. Enjoy your Pi! Raspberry Pi OS comes with pre-installed applications for browsing, media playback, coding, and more. You can further customize it by installing additional software.

Option 2: Flashing the OS yourself:

1. Download Raspberry Pi OS: Head to the official Raspberry Pi website and download the latest version of Raspberry Pi OS Lite (recommended for beginners) or Desktop (full graphical interface). Choose the appropriate image for your Raspberry Pi 5.

2. Flashing tool: Download and install an SD card flashing tool like Etcher or Raspberry Pi Imager. These tools make the process easy and user-friendly.

3. Prepare the microSD card: Insert the card into your computer's SD card reader. Run the flashing tool and select the downloaded image and your microSD card. Follow the tool's instructions to flash the OS onto the card.

4. Safely eject the card: Once complete, safely eject the card from your computer.

5. Insert and power on: Follow steps 2-5 from Option 1 to insert the card into your Pi 5, power it on, and set up your system.

Tips:

- Use a high-quality microSD card (minimum 16GB recommended) for better performance and reliability.
- Choose the Lite version of Raspberry Pi OS if you're comfortable with the command line or plan to use it headless (without a monitor).

- Ensure your computer has a stable internet connection when downloading and flashing the OS.
- Refer to the official Raspberry Pi website for detailed guides and troubleshooting tips if you encounter any issues.

Now that you have your Raspberry Pi 5 setup with Raspberry Pi OS, the possibilities are endless! Start exploring coding, building projects, or simply enjoying the multimedia capabilities. Remember, the journey of learning and creativity is just as exciting as the destination itself. Have fun and discover the magic of Raspberry Pi!

Booting up your Raspberry Pi for the first time

The moment of truth has arrived! Booting up your Raspberry Pi 5 for the first time is a thrilling experience, opening doors to a world of creative possibilities. Here's what to expect:

1. Initial Power-up:

- Connect everything up: Ensure your Raspberry Pi 5 is connected to a monitor via HDMI, keyboard and mouse via USB, and powered with a suitable power supply.
- Lights and LEDs: You'll see various lights and LEDs illuminate on the board, indicating power flow and boot process.

- Rainbow splash: Soon, you'll be greeted by a vibrant rainbow splash on your monitor, followed by a series of white dots. This signifies the bootloader working its magic.

2. Boot Sequence and Console:

- Kernel messages: Once the rainbow sequence completes, you'll see text messages scroll down the screen, displaying the loaded kernel and modules. This might seem cryptic, but it indicates the operating system is initializing.
- Login prompt: Finally, you'll be presented with a login prompt, typically resembling "pi@raspberrypi login:" This

means you're ready to interact with your Pi!

3. Welcome Wizard (Optional):

- Pre-installed OS: If you have a pre-installed microSD card with Raspberry Pi OS, you'll be automatically whisked into the Welcome Wizard. This intuitive interface guides you through setting up your language, Wi-Fi connection, username/password, and time zone.
- Manual setup: If you flashed the OS yourself or chose a headless OS (no graphical interface), you'll need to configure settings manually via the command line. Learn basic Linux commands or

visit the Raspberry Pi website for detailed instructions.

4. Exploring your Pi:

- Desktop environment (Desktop OS only): With Raspberry Pi OS Desktop, you'll land on a familiar-looking desktop with pre-installed applications like web browser, office suite, and media player. Explore, customize, and install additional software to personalize your experience.
- Command line interface: In other OS options or headless setups, you'll utilize the command line to navigate folders, install software, and manage your Pi. Don't worry, beginner-friendly resources are readily available!

Tips:

- Be patient during the boot process, especially on the first boot. It may take a few minutes for everything to initialize.
- If you encounter any issues, check the official Raspberry Pi website for FAQs and troubleshooting guides.
- Remember, the first boot is just the beginning! Enjoy the learning process, experiment with different projects, and embrace the endless possibilities of your Raspberry Pi 5.

So, plug in, power up, and get ready to embark on a thrilling journey of coding, building, and creating with your new Raspberry Pi! The potential is limitless, and the adventure starts now.

CHAPTER THREE

Navigating the Operating System

Here's a guide on navigating the Raspberry Pi OS, covering both desktop and command-line interfaces:

Desktop Environment (Raspberry Pi OS Desktop):

- **Start Menu**: Similar to Windows, the start menu (raspberry icon in the top-left) provides access to applications, settings, and system utilities.

- **Taskbar**: Located at the bottom, it displays open applications and system icons like volume control and network status.

- **File Manager**: Open the "Files" application to browse and manage files and folders, just like on a Windows or macOS computer.

- **Applications**: Launch pre-installed software like Chromium (web browser), LibreOffice (office suite), and VLC Media Player.

- **Customization**: Customize your desktop experience by changing the wallpaper, rearranging icons, and adjusting settings to your preferences.

Command Line Interface (Terminal):

- **Accessing the terminal:**

    - Desktop OS: Open the "Terminal" application from the start menu.

- Other OS versions: You'll start directly in the terminal after booting.

- Basic commands:

　- `ls`: Lists files and directories in the current folder.

　- `cd <directory>`: Changes to a different directory.

　- `pwd`: Displays the current working directory.

　- `sudo <command>`: Executes a command with administrative privileges.

　- `apt-get update`: Updates the list of available software packages.

　- `apt-get install <package>`: Installs a new software package.

- **Resources**:

Explore beginner-friendly tutorials and cheat sheets online for learning more about Linux commands.

Additional Tips:

- **Keyboard shortcuts**: Utilise keyboard shortcuts like Ctrl+C (copy), Ctrl+V (paste), and Ctrl+Alt+T (open terminal) for efficiency.

- **Help**: Access the built-in help system in the terminal with the `man <command>` command.

- **Online resources**: Consult the Raspberry Pi Foundation's website, forums, and online communities for comprehensive guides and troubleshooting.

Remember: Whether you prefer a graphical desktop or the power of the command line, Raspberry Pi OS offers flexibility to suit your preferences and skill level. Embrace the learning process, explore different functionalities, and have fun discovering the capabilities of your Raspberry Pi 5!

Basic user interface elements and functionalities

Here's a breakdown of basic user interface elements and functionalities in Raspberry Pi OS, covering both desktop and command-line interfaces:

Desktop Environment (Raspberry Pi OS Desktop):

Common Elements:

- **Desktop**: The primary workspace where you'll interact with windows and icons.

- **Start Menu**: The central hub for accessing applications, settings, and files (usually a raspberry icon in the top-left corner).

- **Taskbar**: Located at the bottom, it displays open applications, system icons, and the clock.

- **File Manager**: The "Files" application for managing files and folders (similar to Windows Explorer or Finder on macOS).

- **Application Windows:** Individual windows for running programs, usually with resizable borders, close buttons, and menu bars.

Interacting with the Desktop:

- **Point and Click:** Use your mouse or touchpad to navigate and select items.

- **Double-click**: Open files and folders.

- **Right-click**: Access context menus for additional options.

- **Drag and Drop**: Move files and folders between locations.

- **Keyboard Shortcuts**: Use combinations like Ctrl+C (copy), Ctrl+V (paste), and Alt+Tab (switch windows).

Command Line Interface (Terminal):

Basic Elements:

- **Command Prompt:** The line where you type commands.

- **Cursor**: Indicates where you're typing.

- **Output**: Displays command results or system messages.

Interacting with the Terminal:

- **Typing Commands**: Enter instructions for actions like listing files (`ls`), changing directories (`cd`), or installing software (`apt-get install`).

- **Arrow Keys:** Navigate through command history.

- **Tab Key**: Autocomplete commands and filenames.

- **Up/Down Arrows**: Scroll through previous commands and output.

Remember:

- Explore both desktop and command-line interfaces to discover the best ways to interact with your Raspberry Pi based on your preferences and tasks.

- Don't hesitate to use online resources and tutorials for further guidance and troubleshooting.

- Embrace the learning process and experiment with different functionalities to unlock the full potential of your Raspberry Pi!

Managing files and folders

Managing files and folders on your Raspberry Pi 5 is essential for organization and efficiency. Here's how to do it using both desktop and command-line interfaces:

Desktop Environment (Raspberry Pi OS Desktop):

- **File Manager:** Use the "Files" application to visually navigate and manage files and folders.

   - Open, move, copy, paste, create, delete, and rename files and folders using familiar icons and menus.

   - Right-click on items for context menus with additional options.

   - Drag and drop files and folders to move them between locations.

Command Line Interface (Terminal):

- Basic file commands:

- `ls`: Lists files and directories in the current folder.

- `cd <directory>`: Changes to a different directory.

- `pwd`: Displays the current working directory.

- `mkdir <directory>`: Creates a new directory.

- `rm <file>`: Deletes a file.

- `mv <source> <destination>`: Moves or renames files and directories.

- `cp <source> <destination>`: Copies files and directories.

- `nano <file>`: Open a text file for editing.

Additional Tips:

- **File extensions**: Pay attention to file extensions (e.g., .txt, .jpg, .py) to understand file types and compatibility.

- **File permissions**: Linux systems have file permissions for security. Use `chmod` commands to adjust permissions if needed.

- **Searching for files**: Use the `find` command in the terminal for advanced file searches.

- **Backups**: Regularly backup important files to an external drive or cloud storage for safety.

Remember:

- Choose the method (desktop or command line) that best suits your preferences and tasks.

- Explore online resources and tutorials for further guidance on file management in Linux environments.

- Practice and experiment to become comfortable with file operations and organization on your Raspberry Pi.

Connecting to the internet and Wi-Fi

Connecting to the internet and Wi-Fi on your Raspberry Pi 5 is a fairly straightforward process, whether you prefer the graphical interface or the command line. Here's a breakdown of both methods:

Using the Desktop Environment:

1. Open the Start Menu: Look for the raspberry icon in the top-left corner.

2. Select "Network Settings": This brings up your network configurations.

3. Choose your Wi-Fi Network: Click on the name of the Wi-Fi network you want to connect to.

4. Enter the Password: Type the password for your chosen network and click "Connect".

5. Wait for Connection: The Pi will attempt to connect. Look for the network icon in the taskbar to confirm connection.

Using the Command Line:

1. Open the Terminal: You can find it in the Start Menu or use the keyboard shortcut Ctrl+Alt+T.

2. List Wi-Fi Networks: Enter the command `iwlist wlan0 scan` to see available networks.

3. Choose your Network: Identify the network name (ESSID) from the list.

4. Connect to the Network: Enter the command `sudo wpa_passphrase <network_name> > /etc/wpa_supplicant.conf`, replacing `<network_name>` with the actual name you chose.

5. Enter the Password: When prompted, type the password for your chosen network.

6. Restart the Networking Service: Enter the command `sudo systemctl restart networking`.

Additional Tips:

- Make sure your Pi has the latest Raspberry Pi OS image installed for better compatibility and updated drivers.
- If you're using Ethernet instead of Wi-Fi, simply connect the cable to the Ethernet port on your Pi and it should automatically detect the connection.
- If you encounter connection issues, consult the Raspberry Pi website for troubleshooting guides and FAQs.
- Explore advanced networking configurations in the Desktop Environment's network settings

or through commands like `nmcli` for more control over your network behavior.

Remember, Connecting to the internet opens up a world of possibilities for your Raspberry Pi 5. Browse the web, download software, stream content, and utilize countless online resources, all thanks to a stable internet connection. Enjoy exploring the digital world with your connected Pi!

PART TWO

Exploring the Capabilities of Raspberry Pi 5

Multimedia Center and Retro Gaming

Raspberry Pi 5: Creating a Multimedia Center and Retro Gaming Heaven

Your Raspberry Pi 5 packs the punch to transform into both a powerful multimedia center and a nostalgic retro gaming console. Here's how to make it happen:

Multimedia Center:

Software:

Kodi: Open-source media player with a user-friendly interface for playing audio, video, and even streaming services.

Plex Media Server (Optional): Turn your Pi into a central media server, streaming content to other devices on your network.

Emby (Optional): Another popular media server option with additional features like live TV recording and parental controls.

Hardware:

HDMI cable: Connect your Pi to your TV or projector.

USB storage: Store your media files on a USB drive or external hard drive.

Remote control (Optional): Enhance your convenience with a dedicated remote for media playback.

Setup:

1. Install your chosen media center software like Kodi.

2. Add your media files to the Pi's storage or network location.

3. Customize the interface and settings to your preferences.

4. Sit back, relax, and enjoy your movies, music, and shows!

Retro Gaming:

Software:

RetroPie: Popular emulation platform with support for thousands of classic games from various consoles.

Recalbox: Another user-friendly emulation option with a curated library of classic games.

Lakka: Lightweight and performance-focused emulation

platform ideal for older Raspberry Pi models.

Hardware:

Retro game controllers: USB or Bluetooth controllers compatible with your favorite retro systems.

Optional case: Protect your Pi and add a retro aesthetic with a themed case.

Setup:

1. Install your chosen emulation software like RetroPie.

2. Download game ROMs for the supported systems (legally!).

3. Configure your controllers and map buttons to the emulated games.

4. Dive into the nostalgic world of classic gaming!

Additional Tips:

- Explore online resources and forums for detailed setup guides and troubleshooting tips.
- Choose game ROMs from trusted sources to avoid malware.
- Consider a dedicated arcade controller for the ultimate retro gaming experience.
- Experiment with different emulation platforms and configurations to find what works best for you.

Remember: The possibilities are endless when it comes to your Raspberry Pi 5's multimedia and gaming capabilities. Get creative, experiment, and discover a world of entertainment right at your fingertips!

PLAYING MUSIC, MOVIES, AND TV SHOWS

Your Raspberry Pi 5 can transform into a powerful multimedia center, effortlessly handling your music, movies, and TV shows. Here's how to enjoy your favorite content:

Software Options:

Kodi: This open-source media player reigns supreme with its user-friendly interface, diverse media support, and ability to handle audio, video, and even streaming services.

Plex Media Server (Optional): Want to turn your Pi into a central media server? Plex allows streaming your content to other devices on your network, making it an ideal setup for multi-room entertainment.

Emby (Optional): Another popular media server offering additional features like live TV recording and parental controls, suitable for advanced users.

Hardware Setup:

HDMI cable: Connect your Pi to your TV or projector for stunning visuals.

USB storage: Load your media files onto a USB drive or external hard drive for ample storage.

Remote control (Optional): Enhance your comfort with a dedicated remote for seamless playback control.

Playing Music:

Organize your music library: Kodi and other players will display your music by artist, album, genre, and more. Consider using music tagging software for efficient organization.

High-resolution audio: Enjoy lossless audio formats like FLAC and WAV for exceptional sound quality on compatible equipment.

Music streaming services (Kodi add-ons): Integrate services like Spotify or Tidal into Kodi for access to massive streaming libraries (may require subscriptions).

Movies and TV Shows:

Video codecs: Ensure your chosen software supports the video codecs of your movie and TV show files (e.g., MP4, H.264).

Subtitles and closed captions: Enjoy foreign films or enhance accessibility with subtitle and closed caption support.

Streaming services (Kodi add-ons): Access streaming services like Netflix or Hulu for convenient content discovery (may require subscriptions).

Tips and Tricks:

Explore online resources: Find detailed setup guides, troubleshooting tips, and community forums for your chosen software.

Keep your software updated: Regularly update your media player and operating system for improved performance and bug fixes.

Experiment with add-ons: Kodi offers a vast library of add-ons expanding its capabilities, from streaming services to weather channels.

Personalize your experience: Customize the interface layout, themes, and settings to create a media center that reflects your preferences.

Remember: Your Raspberry Pi 5 is a blank canvas for your multimedia dreams. Embrace the flexibility, explore software options, and build a home entertainment system that caters to your tastes. So, grab your popcorn, dim the lights, and prepare to be entertained by the magic of your Pi-powered media center!

Setting up a retro gaming console with emulators

Your Raspberry Pi 5 holds the key to unlocking a treasure trove of nostalgic gaming experiences. Here's how to unleash its retro gaming potential with emulators:

Choosing Your Weapon: Popular Emulation Platforms:

RetroPie: Widely popular and user-friendly, offering pre-configured emulators for numerous consoles, from Atari to PlayStation 1.

Recalbox: Another user-friendly choice, boasting a curated library of classic games and a simple interface.

Lakka: Lightweight and focused on performance, ideal for older Raspberry Pi models and more advanced users.

Installing Your Emulator Platform:

1. Download the latest image for your chosen platform.

2. Flash the image onto a microSD card using a tool like Etcher.

3. Insert the microSD card into your Raspberry Pi 5 and power it on.

4. Follow the on-screen instructions for initial setup and configuration.

Adding Games: Sourcing ROMs Legally:

- Download game ROMs (copies of game files) only from trustworthy

sources to avoid malware and legal issues.
- Consider purchasing or ripping your own physical game discs for legal ROM acquisition.
- Check online resources for guides on legally obtaining ROMs for specific consoles.

Configuring Controllers and Mapping Buttons:

- Connect your preferred controllers (USB or Bluetooth) to your Raspberry Pi.
- Map the controller buttons for each emulated console to match the original game controls.
- Most platforms offer user-friendly interfaces for controller configuration.

Tips and Tricks for Enhanced Retro Gaming:

Scrape game metadata: Automatically download cover art, descriptions, and ratings for your games for a visually appealing library.

Customize the interface: Change themes, backgrounds, and sounds to create a personalized retro gaming experience.

Experiment with shaders: Enhance visuals with graphical filters like CRT scanlines for an authentic retro feel.

Explore online communities: Join forums and discussion groups for troubleshooting, game recommendations, and retro gaming enthusiasts like yourself.

Remember: The retro gaming possibilities with your Raspberry Pi 5 are endless. Choose your weapon (emulator platform), gather your gaming arsenal (ROMs and controllers), and prepare to embark on a nostalgic journey through the history of video games. Have fun, tinker, experiment, and relive the golden age of gaming with your very own Raspberry Pi retro console!

PROGRAMMING WITH RASPBERRY PI 5

Unlocking Your Raspberry Pi 5's Coding Potential: Python and Scratch, Your Gateway to Programming

Your Raspberry Pi 5 is a versatile platform for learning and creating with code. Python and Scratch offer excellent starting points for different levels of experience:

Python:

- **Versatility**: A general-purpose programming language used for web development, data analysis, scientific computing, and more.

- **Readability**: Known for its clear syntax, making it easier to learn and understand.

- **Beginner-friendly**: Plenty of resources and tutorials available online.

- **Integration with Raspberry Pi:** Pre-installed on Raspberry Pi OS and well-suited for hardware interaction.

Scratch:

- **Visual block-based coding:** Drag-and-drop blocks to create programs, making it intuitive for younger learners and those new to coding.

- **Game creation:** Ideal for designing simple games and animations.

- **Foundation for logic**: Teaches basic programming concepts like loops, conditionals, and variables.

- **Available on Raspberry Pi:** Accessible through the "Programming" menu in Raspberry Pi OS.

GETTING STARTED WITH PYTHON:

1. Open a Python interpreter: Type `python3` in the terminal to start coding interactively.

2. Basic syntax: Learn variables, data types (numbers, strings, lists), operators, and control flow (if statements, loops).

3. Online resources: Explore tutorials, books, and interactive courses like those on Codecademy or Coursera.

4. Experiment: Write simple programs to print messages, do

calculations, or interact with user input.

5. Libraries and modules: Expand Python's capabilities with libraries for graphics, games, web development, and more.

Exploring Scratch:

1. Open Scratch: Find it in the "Programming" menu.

2. Interactive tutorials: Learn the basics through the built-in tutorials.

3. Create projects: Drag blocks together to make animations, stories, or simple games.

4. Share and remix: Join the online Scratch community to share projects and learn from others' creations.

5. Transition to text-based coding: The concepts learned in Scratch can be applied to text-based languages like Python later on.

Remember:

- Choose the language that best suits your age and experience level.

- Start with simple projects and gradually increase complexity as you learn.

- Don't be afraid to experiment, make mistakes, and troubleshoot.

- Utilize online resources, tutorials, and communities for guidance and support.

- Embrace the learning process and have fun exploring the world of programming with your Raspberry Pi 5!

Building simple interactive programs and games

Building Fun and Interactive Programs with Your Raspberry Pi 5: A Beginner's Guide

Your Raspberry Pi 5 is a fantastic playground for tinkering with code and creating simple interactive programs and games. Let's explore some beginner-friendly ideas to get you started:

Interactive Programs:

Guessing Game: Write a program that generates a random number and lets the user guess it within a certain number of tries. Implement feedback like "higher" or "lower" to guide the user.

Mad Libs: Create a Mad Libs game where the user inputs words for different parts of speech, then generates a funny story incorporating those words.

Rock-Paper-Scissors: Build a classic Rock-Paper-Scissors game against the computer, with random choices and scorekeeping.

Simple Games:

Space Invaders: Create a basic version of the classic arcade game with moving spaceships, lasers, and scorekeeping. Use libraries like Pygame or pygamezero for graphics and sound.

Pong: Implement the iconic Pong game with paddles bouncing a ball back and forth across the screen.

Memory Game: Design a matching game where players reveal hidden tiles and try to find pairs. Expand difficulty by increasing the number of tiles.

Tips and Resources:

Start small: Break down complex ideas into smaller, achievable steps.

Utilize libraries: Leverage existing libraries like Pygame or Tkinter for easier graphics and user interface creation.

Online resources: Check out tutorials, websites like Learn Python the Hard Way, and books like "Beginning Game

Programming with Pygame Zero" for guidance.

Communities: Join online forums and communities for Raspberry Pi and Python programming to ask questions, share projects, and get inspiration.

Most importantly, have fun! Experiment, play with different ideas, and don't be afraid to make mistakes. The learning process is just as rewarding as the finished product.

Remember, Building interactive programs and games is a fantastic way to learn coding, hone your problem-solving skills, and unleash your creativity. Your Raspberry Pi 5 offers a perfect platform to embark on this exciting journey. So, start coding, build something fun, and enjoy the magic of your personalized creations!

Learning control flow, variables, and functions

Mastering the Building Blocks of Programming with Python on Your Raspberry Pi 5: Control Flow, Variables, and Functions

I'm ready to guide you through these core concepts:

Control Flow:

Conditional Statements (if/else):

   - Make decisions in your code based on conditions.

   - Example: `if age >= 18: print("You can vote") else: print("You are too young to vote")`

Loops (for/while):

- Repeat code blocks multiple times.

- `for i in range(5): print(i)`

- `while temperature > 100: print("Too hot")`

Variables:

Storing Data: Containers for storing different types of data.

- Numbers: `age = 25`

- Text (strings): `name = "Alice"`

- Lists: `shopping_list = ["apples", "bread", "milk"]`

Naming Rules:

- Start with a letter or underscore.

- Use letters, numbers, and underscores.

- Case-sensitive (age is different from Age).

Functions:

Reusable Code Blocks: Define blocks of code that can be called multiple times.

- Structure: `def function_name(parameters): code_to_execute`

- Example: `def greet(name): print("Hello, " + name)`

- Calling: `greet("Bob")`

Benefits:

- Organization: Break down complex tasks into smaller, manageable functions.

- Reusability: Write code once, use it multiple times.

- Modularity: Test and debug functions independently.

Practice!: Experiment with code examples, build small projects, and solve coding challenges.

Remember:

- Start with basic concepts and gradually build complexity.

- Practice regularly to solidify understanding.

- Don't hesitate to seek help from online communities and forums.

- Embrace the learning process and enjoy the satisfaction of mastering new skills!

GLOSSARY OF TERMS

Raspberry Pi 5 Glossary of Key Terms and Abbreviations:

Hardware:

CPU: Central Processing Unit - the brain of the Raspberry Pi, responsible for processing instructions.

GPU: Graphics Processing Unit - handles graphics processing for visuals and video.

RAM: Random Access Memory - temporary storage for data used by currently running programs.

ROM: Read-Only Memory - permanent storage for the operating system and boot files.

GPIO: General-Purpose Input/Output - pins that can be used to connect sensors, buttons, and other electronics.

HDMI: High-Definition Multimedia Interface - for connecting to TVs and monitors.

USB: Universal Serial Bus - standard connection for peripherals like keyboards, mice, and external storage.

MicroSD card: Storage medium for the operating system and files.

Software:

Raspberry Pi OS: Official operating system for Raspberry Pi, based on Debian Linux.

Emulator: Software that simulates another computer system, allowing you to run older games or software.

ROM: Refers to a game file (Read-Only Memory) for retro gaming with emulators.

Python: Popular programming language with beginner-friendly syntax, often used on Raspberry Pi.

Scratch: Visual block-based programming language, ideal for younger learners and beginners.

SSH: Secure Shell - secure remote access protocol for accessing your Pi from another computer.

General:

LED: Light Emitting Diode - small lights used on the Pi for status indicators.

Booting:?))Process of starting up the Raspberry Pi and loading the operating system.

Desktop environment: Graphical interface for interacting with your Pi using windows, icons, and a mouse.

Command line: Text-based interface for interacting with your Pi using typed commands.

IP address: Unique identifier for your Pi on a network.

SSH: Secure Shell - secure remote access protocol for accessing your Pi from another computer.

ABBREVIATIONS:

Pi: Short for Raspberry Pi.

OS: Operating System.

GB: Gigabyte - unit of storage capacity.

MHz/GHz: Megahertz/Gigahertz - units of clock speed for the CPU.

IDE: Integrated Development Environment - software suite for writing and running code.

API: Application Programming Interface - set of tools for interacting with a program or service.

Remember: This is just a starting point. As you explore the world of Raspberry Pi, you'll encounter many more terms and abbreviations. Don't hesitate to look them up or ask for help to expand your knowledge and understanding!

TROUBLESHOOTING TIPS AND FAQS

Raspberry Pi 5 Troubleshooting Tips and FAQs: Smooth Sailing on Your Tech Voyage!

Even the most seasoned Raspberry Pi enthusiasts encounter hiccups occasionally. Don't fret, here's a toolkit to help you diagnose and fix common issues:

General Troubleshooting:

Problem: No Power

Solution:

* Check the power adapter connection (both to the wall outlet and Pi).

* Try a different power adapter or cable.

* Ensure the microSD card is properly inserted.

Problem: No Boot

Solution:

* Confirm the microSD card has the correct operating system image flashed.

* Re-flash the operating system image to the microSD card.

Check for damaged components on the Pi.

Problem: Overheating

Solution:

Improve ventilation around the Pi.

Add heatsinks or fans to the CPU and GPU.

Check for dust buildup and clean vents.

Problem: Network Issues

Solution:

Verify Wi-Fi is enabled and configured correctly.

* Check network cable connection (if using Ethernet).

* Restart networking service using the command `sudo systemctl restart networking`.

*Software Issues:

Problem: Corrupted System

Solution:

- Re-flash the operating system image to the microSD card.
- Backup important files before performing any actions.

**Problem: Application not working

Solution:

Update the application to the latest version.

Check for dependencies and install any missing ones.

Consult the application's documentation or online forums for specific troubleshooting steps.

FAQs:

Q: Can I upgrade the RAM on my Raspberry Pi 5?

A: Unfortunately, the RAM on the Raspberry Pi 5 is soldered onto the board and cannot be upgraded.

Q: What are some good resources for learning about Raspberry Pi?

A: Official Raspberry Pi website, Raspberry Pi forums, online tutorials from trusted sources like Codecademy, books like "Raspberry Pi User Guide" or "Automate the Boring Stuff with Python".

Q: Where can I find projects and ideas for my Raspberry Pi?

A: Online communities like Reddit's r/raspberry_pi, websites like Hackster.io, or project books like "Make Projects with Raspberry Pi".

Remember: Don't hesitate to reach out to online communities and forums for specific troubleshooting assistance. There's a vast and helpful Raspberry Pi community out there to guide you through any challenges!

Enjoy your Raspberry Pi 5 adventures, and keep exploring the endless possibilities!

www.ingramcontent.com/pod-product-compliance
Lightning Source LLC
LaVergne TN
LVHW010019070125
800679LV00029B/1405